This book belongs to:

____

Made in the USA
Columbia, SC
19 December 2022

74518439R00067